Animals with Venom
Blue-Ringed Octopus
by Julie Murray

Dash!
LEVELED READERS
An Imprint of Abdo Zoom • abdobooks.com
1

Dash!
LEVELED READERS

Level 1 – Beginning
Short and simple sentences with familiar words or patterns for children who are beginning to understand how letters and sounds go together.

Level 2 – Emerging
Longer words and sentences with more complex language patterns for readers who are practicing common words and letter sounds.

Level 3 – Transitional
More developed language and vocabulary for readers who are becoming more independent.

THIS BOOK CONTAINS RECYCLED MATERIALS

abdobooks.com

Published by Abdo Zoom, a division of ABDO, PO Box 398166, Minneapolis, Minnesota 55439. Copyright © 2021 by Abdo Consulting Group, Inc. International copyrights reserved in all countries. No part of this book may be reproduced in any form without written permission from the publisher. Dash!™ is a trademark and logo of Abdo Zoom.

Printed in the United States of America, North Mankato, Minnesota.
052020
092020

Photo Credits: Alamy, iStock, Minden Pictures, Shutterstock
Production Contributors: Kenny Abdo, Jennie Forsberg, Grace Hansen, John Hansen
Design Contributors: Dorothy Toth, Neil Klinepier, Candice Keimig

Library of Congress Control Number: 2019956194

Publisher's Cataloging in Publication Data
Names: Murray, Julie, author.
Title: Blue-ringed octopus / by Julie Murray
Description: Minneapolis, Minnesota : Abdo Zoom, 2021 | Series: Animals with venom | Includes online resources and index.
Identifiers: ISBN 9781098221027 (lib. bdg.) | ISBN 9781644943977 (pbk.) | ISBN 9781098222000 (ebook) | ISBN 9781098222499 (Read-to-Me ebook)
Subjects: LCSH: Blue-ringed octopuses--Juvenile literature. | Mollusks--Juvenile literature. | Poisonous animals--Juvenile literature. | Poisonous marine animals--Juvenile literature. | Bites and stings--Juvenile literature.
Classification: DDC 591.69--dc23

Table of Contents

Blue-Ringed Octopus 4

More Facts 22

Glossary 23

Index 24

Online Resources 24

Blue-Ringed Octopus

The blue-ringed octopus lives in the Pacific and Indian Oceans.

It spends its life in shallow **tide pools**. It also lives in coral reefs.

The blue-ringed octopus is small. It is less than 8 inches (20 cm) long!

The blue-ringed octopus is mostly yellow in color. It is named for its blue rings that glow when it is angry.

Its bite is often painless. But its **venom** is deadly!

The blue-ringed octopus has eight arms. The arms are covered in suction cups. These help the octopus hold onto things.

The blue-ringed octopus has two eyes. Its **beak** helps it tear food apart.

17

Its **beak** also helps the octopus eat shelled animals. It likes crabs and lobsters!

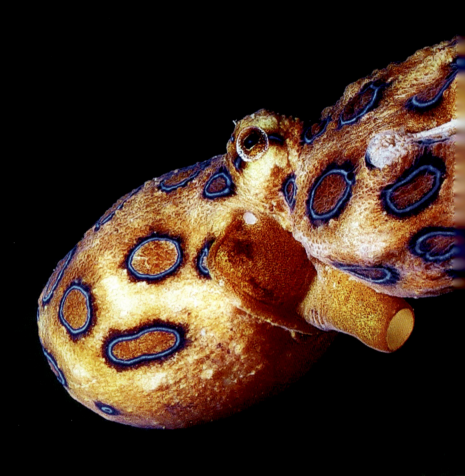

Like all octopuses, they do not have bones. They are **flexible**. They can fit into tight spaces.

More Facts

- A blue-ringed octopus has enough **venom** to kill 25 people!

- A blue-ringed octopus's bite causes pain and numbness. The venom can affect the **respiratory system**, which is when it can become deadly.

- It is not an **aggressive** animal. It only bites if handled, stepped on, or threatened.

Glossary

aggressive – unfriendly, forceful, and ready to fight.

beak – the hard, curved mouth part of some animals.

flexible – easily bent without breaking.

respiratory system – the organs and other parts of the body involved in breathing.

tide pool – a body of water from the sea that remains when the waves move back from the shore. Many types of animals and plants can live in a tide pool.

venom – the poison that certain animals make.

Index

arms 14

beak 17, 19

bite 12

color 11

eyes 17

food 17, 19

habitat 4, 7

Indian Ocean 4

Pacific Ocean 4

size 9

suction cups 14

venom 12

Online Resources

Booklinks
NONFICTION NETWORK
FREE! ONLINE NONFICTION RESOURCES

To learn more about the blue-ringed octopus, please visit **abdobooklinks.com** or scan this QR code. These links are routinely monitored and updated to provide the most current information available.